W9-AGK-992

3 1489 00434 4725

Booklet

3 1489 00431 3571

METRIC FOR ME !

A Layperson's Guide To
THE METRIC SYSTEM FOR EVERYDAY USE
With
Exercises, Problems, and Estimations

2nd Edition, Enlarged & Revised

by **ROBERT W. SHOEMAKER**
Professor Emeritus of History
North Central College

Blackhawk Metric Supply Inc.
South Beloit, Illinois
1998

© Copyright 1993, 1998 by Robert W. Shoemaker

All rights reserved. Printed in the United States of America.
No part of this book may be reproduced or transmitted in any form or by
any means, electronic or mechanical, including photocopying, recording, or
by any information storage and retrieval system without written permission
from the author, except for brief quotations in critical articles and reviews.
    For information contact Blackhawk Metric Supply Inc., PO Box 543,
South Beloit IL 61080.

 blackhawk metric supply, inc.
p.o. box 543
south beloit, il 61080

Library of Congress Catalog Card Number 96-79294

ISBN 0-9627980-3-7

10 9 8 7 6 5 4 3 2 1

For my very dear Grandchildren

Jesse

Anna Leigh

Paul Robert

And in loving Memory of

My precious wife Caroline

WHY METRIC ?

The United States has about 5% of the world's population. Most of the other 95% of the people in the world use the metric system of measurement.

Since 1980, the American foreign exchange deficit has risen enormously, standing over $114 billion in 1996.

American goods manufactured in inches and pounds are less welcomed in the metric world market than are goods of comparable quality produced in metric by foreign companies.

And as American sales are lost to foreign competitors, so too are American jobs.

Although changing to metric will not eliminate all of America's foreign trade problems, that change will certainly make American goods more saleable abroad, thereby helping to reduce our tremendous foreign trade deficit and preserving our jobs.

The European Union of 15 nations with uniform trade requirements is the world's largest economic unit containing 370 million people with an annual spending power of over $7.5 trillion. And because of its growing body of metric-based regulations, the EU—the world's richest market—will increasingly demand metric-only goods and services.

Clearly, the United States *cannot afford* to continue manufacturing and shipping in feet and pounds. It *must* adopt the world system of metric units. Therefore, in the Omnibus Trade and Competitiveness Act of 1988, the United States Government stated its intention to do just that.

Come join those American companies and people that are changing to metric! Not only will you find that metric makes good economic sense, but also you will discover why the rest of the world has been using it for so long:

— because metric is just so much more logical, easy to learn and simpler to use than inches and ounces, pecks and pounds!

TO THE READER

INTERNATIONAL SYSTEM (SI) METRIC

AS THE TITLE says, this is a *Layperson's Guide to the Metric System for Everyday Use*. Its intended audience is that very large number of people who are not scientists, people with all sorts of different abilities, interests, and experiences. Such an audience is difficult to write for because some of its members will find parts of this book too simple while others will find parts of it too hard. It is impossible to write a book that is just right for each one of these laypeople. So when or if you come to a portion that seems overly easy or difficult, please just forgive the author and keep on reading.

Closely related to this book is the accompanying chart entitled *All You Need To Know About The Metric System For Everyday Use*. This is a helpful companion to the text (which will frequently refer to the chart) because ideas presented in written form in the book are seen again and in a different manner in colored, pictorial form in the chart. And while the book benefits by having its ideas presented pictorially in the chart, the chart does not need the book to present its ideas. The chart is complete in itself and thus can be purchased separately.

Your attention is called to the ample *Hands-on Exercises, Review Problems,* and *Estimations* which follow each major topic (i.e., Length, Volume, and Weight) in the book. These exercises, problems, and estimations give you many opportunities to put into practice the metric concepts about which you have just read and to understand them well. Pay close attention as you work through these sections; such conscientious effort will benefit your learning very much.

An unusual aspect of this book is the appearance of blank pages at various intervals, namely, after each set of Estimations. These blank pages are to be used for whatever notes you may want to make regarding the major topics, the hands-on work, or any calculations for the problems. Thus you do not need to bring a notebook or loose paper to class with you. Everything relative to your learning metric will be contained in this one book. You will not be bothered by having information scattered here and there in various papers.

If you are using this book by yourself rather than as a member of a class, you will find another noteworthy aspect, namely, that the book and the accompanying chart present you with 1) A full course on metric, 2) A short course, 3) A very short course, or 4) Just a quick overview. (If you are in a class, the instructor will probably have predetermined the length of the course.)

For 1) The full course, use the entire book and the chart. For 2) The short course, use the chart but omit everything in the book up to page 6, THE METRIC SYSTEM: HOW TO USE IT; then skip all the Hands-on Exercises, Review Problems, and Estimations; continue through page 49, TEMPERATURE, and stop there. For 3) The very short course, use the chart alone. Or, for 4) A quick overview, just read page 52 in the book, SUM-MARY OF METRIC.

As you read this book you will discover the importance of measuring things in *metric units only* without reference to such English units as feet or ounces. But in usual retail outlets in the United States, whether stores or catalogs, it will be practically impossible to find metric-only measuring tapes, scales, or other such devices. However, in the catalogs of educational supply companies you will find just such equipment. To see these catalogs, simply ask the person in charge of mathematics or science at a nearby grammar or high school who, quite likely, will allow you to examine the catalogs in his/herpossession. In these catalogs you will find a good variety of metric-only measuring equipment.

As laypeople, you are perhaps not familiar with some different names for the metric system which you will encounter such as "International System," or "SI," or "SI Metric." So right here we should explain their meaning and significance by looking at a bit of the history of the metric system of measurement:

The multi-national Treaty of the Meter of 1875, of which the United States was an original signer, created the International Bureau of Weights and Measures (located outside Paris) along with an international General Conference on Weights and Measures to supervise use of the metric system (which had been introduced in the 1790s) and to continually refine its accuracy in accordance with the latest pertinent scientific developments. As a result of the work of the Bureau and the Conferences (held approximately every four years), the degree of accuracy and reproducibility of metric measuring units had been improved so much that by 1960 a "new" metric system based on standards of natural phenomena different from those used in the 1790s had been created. So the eleventh international Conference named this modernized metric system the "Système International d'Unités" or "International System of Units." Known universally by its French abbreviation of "SI," this new metric system has, itself, been constantly improved upon since 1960 and serves as an ever more precise and reliable system of measurement.

The course on the metric system presented in this book is for the layperson, not for the scientist who already knows metric anyway. Thus in situations where subsequent scientific refinements have led to discrepancies between SI and traditional metric use, we will follow whichever practice seems better suited to the layperson. For instance, we will always consider a liter of water as weighing the traditional 1 kilogram, not the precise SI

1.000,028 kilograms. The aim of this course, after all, is clarity of concepts, insight into the logic of metric, ease of understanding, ease of using, and ease of retaining a knowledge of the metric system for everyday use.

In closing, the author wishes to express his deep gratitude to a man who contributed so much to helping the United States enter the metric world and who assured publication of the accompanying metric chart. This helpful and generous man, who regrettably died in 1994, was Mr. Kenyon Y. Taylor, CEO of North American Tool Corporation in South Beloit, Illinois.

The author also wishes to express his appreciation for the important help which he received from the metric experts who read and criticized the MS of this work before publication. These well-informed and helpful people were Mrs. Valerie Antoine, Executive Director, and Mrs. Lorelle Young, President, of the U.S. Metric Association; Dr. David T. Goldman, past Secretary of the Consultative Committee on Units of the International Committee for Weights and Measures; and Mr.Gerald T. Underwood, past Director of the Office of Metric Programs, U.S. Department of Commerce and former President of the American National Metric Council. Many of the helpful suggestions of these knowledgeable people were utilized in the final revision of this work, but any remaining shortcomings or errors are the sole responsibility of the author.

And finally, for additional information about metric or for information about membership in a metric organization, contact any of the following:

U.S. Metric Association,
10245 Andasol Av.,
Northridge CA 91325;
Tel: 818-368-7443,
FAX: 818-368-7443

American National Metric Council,
4340 East-West Highway, Suite 401,
Bethesda MD 20814;
Tel: 301-718-6508,
FAX: 301-656-0989,
e-mail: ANMC@paimgmt.com

Metric Program Office,
National Institute of Standards and Technology,
U.S. Department of Commerce,
Building 820, Room 306,
Gaithersburg MD 20899;
Tel: 301-975-3690,
FAX: 301-948-1416,
e-mail: metric_prg@nist.gov

The USMA is essentially a group of individual people with some business firms as members, and the ANMC consists mostly of companies with some individuals as members.

Naperville IL, March 1993 (updated January 1998)   — RWS

PREFACE TO THE SECOND EDITION

It is always gratifying to both author and publisher when a book has been so well received that a second edition is called for. Both author and publisher thank the readers for showing such confidence in this book, and we hope that even more readers will benefit from the additions and revisions made in this edition.

The major difference between this and the first edition is the addition of the Technological Supplement; then minor differences are to be found here and there in the revisions.

To deal first with the minor differences: Even in the few years since the appearance of the first edition, some conditions have changed, so the text has been updated in sundry places to reflect these changes. Also, in the Timetables for Courses of various lengths, the eight hour timetable has been eliminated in the light of classroom experience.

But the Technological Supplement is the greatest difference. As originally written, this book was solely for laypeople, including elementary school children. However, many buyers of the book were people in the highway construction, the building construction, and the electric power industries. To better serve such people as well as those in other technical fields, the Technological Supplement has been added with the hope that it will substantially increase the usefulness of *Metric For Me!*

In closing, the author wants to thank Dr. Gary P. Carver, past Director of the U.S. Department of Commerce Metric Program, for his helpful reading and criticism of the revisions in this edition. And finally, both author and publisher would appreciate readers' comments regarding this second edition so that any future edition may even better serve the public.

Naperville IL, January 1998   — RWS

TABLE OF CONTENTS

TIMETABLES FOR 2, 4, and 6 HOUR COURSES

ONE CONVENIENCE of this book is that it is suitable for courses of different lengths and it presents specific timetables for courses of 2, 4, or 6 hours' duration. Furthermore, the instructor can make even shorter or longer versions as desired.

The 6 hour course is the most desirable one because it presents the most information and instruction and is thus the standard upon which the other two courses are based. But as a practical matter, sitting for six hours of class work may be pretty wearisome, and even two hours can be tiring. Therefore the 6 hour course is divided into two sessions, perhaps one for today and one for another day, and within the sessions convenient break points are designated. By taking advantage of these break points, a person can even spread the course over several days if desired. Or, just the opposite, another person might need a concentrated 6 hour course in one day, about three hours in the morning and three in the afternoon. Such a course is available by using the 6 hour timetable as it stands.

Even the short courses, the 2 and 4 hour ones, can be presented as complete two or four hour units at one time or can be broken down. By using the break points, the 4 hour course can be divided into three sessions and the 2 hour course into two.

In short, *despite the differing time needs of different offices, companies, schools, or other groups*, this book as an instructional medium will fit the requirements of practically everyone. It attains its flexibility, of course, by providing for the systematic deletion of material, especially "reinforcing" material, i.e., material in the sections of the book called Hands-on Exercises, Review Problems, and Estimations. Yet some exercises and problems deal with concepts of such importance that these specific exercises and problems must be retained. In the Hands-on and Problem portions you will find them conveniently singled out by the symbol •.

The timetables for the various courses follow, beginning with the complete 6 hour course. The subject headings listed in the 6 hour course correspond to the subject headings in the Table of Contents of this book. The same is true in the 4 and 2 hour course timetables where you will notice that more and more subject headings are deleted.

———————————

NB: The first edition of this book had a timetable for an 8 hour course also. However, actual classroom experience showed that there was not enough material to justify 8 hours, so that timetable was eliminated for this edition.

TIMETABLE OF 6 HOUR COURSE
AND TOPICS COVERED

— First Session —

		Cumulative Total
THE METRIC SYSTEM: BACKGROUND THE METRIC SYSTEM IN THE UNITED STATES }	20 min ——— 20	0 hr 20 min

THE METRIC SYSTEM: HOW TO USE IT
    Introduction }
    Length
	45 min	

Break	10	1 hr 15 min
Hands-on Exercises on Length	35 min	
Review Problems on Length	20	
Estimations on Length	30 ——— 1 hr 25 min	

Duration of First Session 2 hr 40 min

— Second Session —

		Cumulative Total
Volume	20 min	
Hands-on Exercises on Volume	30	
Review Problems on Volume	15	
Estimations on Volume	15 ——— 1 hr 20 min	
Break	10	1 hr 30 min
Mass (Weight)	20 min	
Hands-on Exercises on Weight	30	
Review Problems on Weight	15	
Estimations on Weight	15 ——— 1 hr 20 min	
Break	10	3 hr 0 min
TEMPERATURE	10 min	
Discussion Summarizing Course	10 ——— 20 min	

Duration of Second Session 3 hr 20 min

Total Duration of Course   6 hr 0 min

TIMETABLE OF 4 HOUR COURSE
AND TOPICS COVERED

Cumulative total

THE METRIC SYSTEM: BACKGROUND THE METRIC SYSTEM IN THE UNITED STATES }	10 min	
	10	0 hr 10 min
THE METRIC SYSTEM: HOW TO USE IT		
Introduction Length }	40 min	
Hands-on Exercises on Length	25	
Review Problems on Length	10	
	1 hr 15 min	1 hr 25 min
Break	10	1 hr 35 min
Volume	20 min	
Hands-on Exercises on Volume	25	
Review Problems on Volume	10	
	55 min	2 hr 30 min
Break	10	2 hr 40 min
Mass (Weight)	20 min	
Hands-on Exercises on Weight	25	
Review Problems on Weight	10	
	55 min	3 hr 35 min
TEMPERATURE	10 min	
Discussion Summarizing Course	15	
	25 min	4 hr 0 min
Total Duration of Course		4 hr 0 min

TIMETABLE OF 2 HOUR COURSE
AND TOPICS COVERED

Cumulative total

THE METRIC SYSTEM: HOW TO USE IT

Introduction ⎱		
Length ⎰	35 min	
Hands-on Exercises on Length	10	
Review Problems on Length	5	
	50 min	0 hr 50 min
Break	10	1 hr 0 min
Volume	20 min	
Hands-on Exercises on Volume	5	
Review Problems on Volume	5	
	30 min	1 hr 30 min
Mass (Weight)	20 min	
Hands-on Exercises on Weight	10	
	30 min	2 hr 0 min
Total Duration of Course		2 hr 0 min

THE METRIC SYSTEM: BACKGROUND

The objective of this section is to show
the drawbacks of early measuring systems,
why metric was created, and why it is
superior to all other measuring systems.

EARLY MEASURING SYSTEMS

FROM EARLIEST DAYS, mankind used measuring systems based on a person's body. The Old Testament tells of using a cubit, or the length of a person's forearm, and in Roman days we get to the foot and the mile or "mille passus" of 1000 paces. Although Roman measurement was generally sensible and was retained in use throughout Europe even after the end of the Roman Empire, this measuring system had basic drawbacks. The length of a foot, of course, differed from person to person and, to make matters worse, the Roman foot was divided sometimes into twelve inches and other times into ten. So how long was an inch? Besides that, standardization of weights and measures was neither accurate nor consistent. An ounce according to the standard of the weighmaster in one medieval European city would be heavier (or lighter) than the standard ounce in another city. In fact, in the hundreds of years since the end of the Roman Empire, a great many different measuring units and sizes had appeared in Europe. One result was that by the late 1700s France had "more than 1000 units of measurement...in Paris and the provinces, with approximately 250,000 local variations."[1]

THE METRIC SYSTEM

In this confusing environment where a person did not know just how much a merchant (even an honest merchant) would hand over in the sale of "four feet" of cloth or "five pounds" of flour, there was the obvious need for a logical, simple, and consistent measuring system, one which did not

---

[1] Ronald Edward Zupko, <u>Revolution in Measurement: Western European Weights and Measures Since the Age of Science</u> (Philadelphia: American Philosophical Society, 1990). p 113.

change its sizes from place to place. The result was that, during the French revolution of the 1790s, just such a measuring system was devised, the metric system. Essentially, the metric system consists of only one unit—the base unit called the *meter* (from the Greek word for "measure")—which is used not only to measure length but which also serves as the basis for deriving other units such as those for volume (capacity) and mass (or "weight").

As a logical basic measuring unit, one *which would not change* in size from person to person, and one which presumably could be duplicated accurately by surveyors any place in the world, the French chose the Earth itself as the standard for length and simply defined a *meter as one ten-millionth the distance from the Equator to the North Pole*. For conveniently measuring shorter or longer things, the meter is then divided into smaller parts in steps of ten or multiplied into longer lengths by tens. In order to go from one metric unit to another, all a person has to do is move the decimal point to the left or right rather than having to remember such odd numbers as 12 inches in a foot, 3 feet in a yard, 40 rods in a furlong, and so on. A kilometer composed of exactly 1000 meters certainly makes more sense than a mile of 5280 feet or 1760 yards!

Of course new ideas, even good ideas, run into opposition. People just don't want to change from the old, so the metric system met resistance even in France where it was created and was not finally adopted in that country until 1840. In other lands the metric system was gradually adopted as standard—1849 in Spain, for example, 1862 in Mexico, 1872 in Germany, 1918 in Russia, and so on—until today the entire world uses or is well on the way to using the metric system except for Liberia, Burma, and the United States! But now even the United States is beginning to change, and that is why this course is important to you.

THE METRIC SYSTEM IN THE UNITED STATES

The objective of this section is to show that the United States has some previous experience with metric; the Federal Government is now turning to metric; and therefore, to do their jobs, Government employees and employees of companies doing Government-related work must know the metric system.

EARLY APPEARANCE OF METRIC

ALTHOUGH THE METRIC system is not the customary system of measurement in the United States today, it does have some background in this country. In 1821 John Quincy Adams, Secretary of State and later President of the United States, submitted to the Senate his excellent *Report Upon Weights and Measures* in which he wrote favorably about the then-new metric system. But the United States, accustomed to over 200 years of English weights and measures, and trading more with England than with any other nation, didn't bother to change. Then an act of Congress passed in 1866 declared the metric system legal for use in the United States, and in 1875 the United States joined major nations of the world in signing the Treaty of the Meter which created the International Bureau of Weights and Measures in Paris. This agency is responsible for on-going supervision of the metric system and for making it more accurate in the light of the most recent scientific developments. One result of this treaty is that (since there is no scientific basis for feet, pounds, or other English units of measure) the United States defines feet, pounds, and all other measuring units in metric terms. Thus, for example, the American legal definition of a foot is 0.304,800 meter and that of a pound is 0.453,592 kilogram.

ATTEMPTS TO CHANGE TO METRIC

Scientists in the first half of the 19th century began using the metric system because of its logic and simplicity. Before the year 1900 metric was in use by scientists everywhere including the United States. Other Americans, however, continued using English measurement, and only after World War II did Americans really begin to realize that they were ever more out of step with the rest of the world which already was, or was going, metric. Therefore a 1968 act of Congress directed the Secretary of Commerce to make a study determining the advantages and disadvantages of increased use of metric in the United States. The result, published by the National Bureau of Standards in 1971 and entitled *A Metric America: A Decision Whose Time*

Has Come, recommended that the U.S. change to metric. Consequently, the Metric Conversion Act was passed in 1975, but this law was very ineffective and little was accomplished.

In the meantime, however, many large international U.S. companies changed to metric in their own interest, among them such firms as IBM, General Motors, and John Deere. As an engineer with one of those companies put it, "If you design a machine in the U.S., manufacture it in Germany, sell it in Japan, and repair it with parts made in Mexico, it had better be in metric!" Today the entire American automobile industry is metric, as are pharmaceuticals, computers, medical equipment, and large portions of the chemical, the heavy construction, and the farm machinery industries. In the government realm, building and highway construction is mostly metric; many new NASA programs are metric; the global positioning satellite system is metric; and most new maps issued by Federal Government agencies use metric distances, altitudes, and water depths. Then the Department of Defense, needing interoperability with weapons systems from other countries, has been prominent in Federal Government metric changeover.

However, for a variety of reasons, the United States has fallen behind in world trade, and our foreign trade deficit has risen enormously, standing at $105 billion in 1995. Foreign countries, practically all of which use metric, are ever less eager to buy non-metric American goods, especially when goods of comparable quality can be bought from metric producers in other lands. While nobody claims that merely changing to metric will eliminate all of America's foreign trade problems, the fact is that changing to metric will make American products more welcomed and saleable abroad and will certainly contribute to reducing America's serious trade deficit.

As if uncoordinated international resistance to non-metric American goods were not enough, a coordinated center of resistance emerged at the beginning of 1993. At that time the all-metric European Community of 12 nations (now the European Union of 15 nations) presented the world with uniform trade requirements for the EU's united economic region of some 370 million people in 1995 with a spending power of over $7.5 trillion a year. Because of its ever-growing body of metric-based regulations, the European Union—the world's largest and richest single market—will increasingly restrict the importation of non-metric goods. Unless the United States wants to see its foreign trade diminish more, it had better start producing and shipping in metric!

TRADE ACT OF 1988

In view of the deterioration of American foreign trade, the Omnibus Trade and Competitiveness Act of 1988 (Public Law 100-418) in Section 5164 designates metric as "the preferred system...for United States trade and commerce" and "require[s] that each Federal agency...to the extent economically feasible by the end of the fiscal year 1992 use the metric system...in its procurements, grants, and other business related activities, except to the extent that such use is impractical or is likely to cause...loss of markets to United States firms...."

Although the Federal Government is only slowly following its own advice on metric use, U.S. industry is moving more quickly to position itself in the global marketplace. Grocery products are labeled in metric along with English units and some products are dispensed in rational metric sizes such as 250 milliliters of fruit juice or 2 liters of soda pop.

So, many people including yourself will likely find metric being used more and more on your job which could become entirely metric. And you'll find that you will like metric because it's so logical and so much easier to use than English measure. (Quick: What's the difference in quantity between a quart of milk and a quart of strawberries? Between 16 ounces of milk and 16 ounces of cottage cheese?) In fact, you'll wonder why the United States hasn't been using metric long before this. However, better late than never, and this course will help you enter the sensible and convenient world of metric.

THE METRIC SYSTEM: HOW TO USE IT

INTRODUCTION

The objective of this section is to show
the logic of metric and how to learn metric.

THE METRIC SYSTEM, as noted earlier, was deliberately developed as a much more logical and easy system of measurement than any other in existence. And that is the reason why practically all nations and peoples of the world have adopted it and why the United States Government is adopting it now. If you are a Government employee, are otherwise involved in Government-related work, or are engaged in trade or commerce, you too are about to enjoy the simplicity and sensibility of measuring in metric.

Metric *for everyday use* is all that you will learn in this course because that is what you will encounter most. If you need to know any specialized use of metric, you can learn that when the time comes. For example, everyone using metric needs to know what a "meter" and a "gram" are, but only X-ray technicians and related personnel need to know what radiation units like a "gray" or a "sievert" are. *All You Need To Know About The Metric System For Everyday Use* as taught in this course is very conveniently summarized and shown to you in pictorial form in the chart of that same name which accompanies this handbook. Study that chart carefully and see the logical relations between metric units; the chart will be very helpful to you.

Now, one important word of advice before starting to learn the metric system: Don't convert! *Don't convert* meters to feet or kilograms to pounds and vice-versa. Conversion is simply a nuisance. It necessitates cumbersome arithmetic and gives awkward numbers for answers. "Five feet seven and three-eighths inches times what gives meters?" may be important to a craftsman working with old drawings, but it is not part of learning how to use metric.

To learn the metric system, just take a metric ruler or a metric scale and measure things with it as you will do during the Hands-on Exercises in this course. Yet some people will complain in anxiety, "But if I don't know how many inches it is, I won't know how big it is! I must convert." No, this is not true. This is an emotional fear based on the unspoken assumption that God, for some reason, decreed feet, bushels, and ounces as mankind's true units of measure.

Of course, this is not so. And now that you recognize the irrational basis of your fear, quite possibly the fear itself will vanish. Moreover, you know that you learned the English system of yards, inches, quarts (four kinds), ounces (four kinds), pecks (two kinds), and what-not by using them and without converting from anything else. So if you could learn such a wild system as this by simply using it, you can certainly learn the simple and logical metric system the same way!

Let us now turn to learning the metric system itself. Metric, like any measuring system, deals fundamentally with three elements—Length, Volume (or "the capacity of something," such as a box), and Mass (or Weight)—and we will study each of these in order.

LENGTH

The objective of this section is to give students an understanding of metric units of length as well as of metric prefixes and symbols (short forms).

As stated at the beginning of this manual, the unit of length in the metric system—and the basic unit of the whole system—is the *meter.* The meter, in turn, was defined by its originators in the 1790s in a very logical manner as one ten-millionth (1/10,000,000) the distance on the Earth's surface from the Equator to the North Pole. In fact, though, it was later discovered that because of some imprecision in surveying, there was an error in the size of the standard meter. But the error was very small, only about one part in five thousand, and for everyday use did not matter. Also, since the days when the meter was first introduced it has been re-defined and re-calculated several times with ever greater scientific precision until now (i.e., since 1983) it is defined in terms of an absolutely unchangeable standard, namely, the speed of light.[2]

However, for everyday use the above refinements are insignificant and it is still proper to think of a meter as the sensible one ten-millionth the distance from the Equator to the North Pole. But having said that, who of us can visualize such a distance? So try this: a meter is approximately the distance from a man's shoulder to the end of his opposite outstretched arm. (Think of a traffic cop with his arm outstretched to direct the cars.) Also, a

---

[2] In 1983 the 17th General Conference on Weights and Measures which, by the 1875 Treaty of the Meter, is responsible for establishing international metric standards, defined the meter as the distance traveled by light in a vacuum in a period of 1/299,792,458 of a second.

meter is approximately the distance from the floor to the belt buckle of an average man. More specifically, a meter is the length of the meter stick or tape that you will soon be using in this course. Look at the meter stick and become familiar with it so that you impress on your mind just how long a meter is.

Meters

As you can see from handling and using a meter stick, a meter is of a size convenient for measuring something as big as a room, the height of a building, or the length of a ship. Longer things are measured in kilometers, and shorter things in centimeters or millimeters, all of which are related to a meter by some multiple of 10. Thus:

$$
\begin{aligned}
10 \text{ millimeters} &= 1 \text{ centimeter} \\
\left.\begin{array}{l} 100 \text{ centimeters or} \\ 1000 \text{ millimeters} \end{array}\right\} &= 1 \text{ meter} \\
1000 \text{ meters} &= 1 \text{ kilometer}
\end{aligned}
$$

Do you see how easy it is to do everything by tens rather than having fractions (say, 32nds) of an inch, then 12 inches to a foot, 5280 feet to a mile? Notice that to change units in metric, all you do is move the decimal point to the right or the left, thus:

$$
\begin{aligned}
1{,}259{,}678 \quad \text{millimeters} &= \\
125{,}967.8 \quad \text{centimeters} &= \\
1{,}259.678 \quad \text{meters} &= \\
1.259{,}678 \quad \text{kilometers}
\end{aligned}
$$

Now start with 1,259,678 inches and try to go quickly to feet, rods, furlongs, and miles!

Centimeters and Millimeters

Turning our attention to the shorter metric units of length, we find that centimeters (one hundred of these on your meter stick) are convenient for measuring things approximately the size of a person. An ordinary man might be about 175 centimeters tall, his waist 96 centimeters around, his desk 76 centimeters high, and a book on that desk 16 centimeters wide by 24 high. Very small things are measured in millimeters (1/10 of a centimeter). The small letters in the above-mentioned book are 2 millimeters high and the capital letters 3 millimeters; an American dime is close to 1 millimeter thick, a nickel is about 2 millimeters thick, and a pencil about 8 millimeters thick.

Of course, in the metric system we could just as well say that a man is 1.75 meters tall, the desk 0.76 meters high, and the pencil 0.008 meters thick. Or, considering these sizes another way, we could say the man is 1750 millimeters tall, the desk 760 millimeters high, and the pencil 8 millimeters thick. All of these are correct and are meaningful. However, we generally use millimeters for small things, centimeters for things about the size of a person, and meters for larger things.[3]

Kilometers

Then for longer distances we use kilometers, units of 1000 meters each. To understand how long a kilometer is, look at the chart *All You Need To Know About The Metric System For Everyday Use* and you will see a picture of one kilometer on Chicago's lakefront. And, if you are familiar with Chicago, you will recognize that this kilometer is the distance from the front of the north steps of the Field Museum to the center line of Buckingham Fountain. For familiar one-kilometer stretches in other major American cities, consult the list on the next two pages. Distances within a city (except for short ones) are measured in kilometers, as are distances between cities. Thus the straight-line distance from Chicago to Washington is just about 1000 kilometers; that from Chicago to New York is virtually 1200 kilometers, and Chicago to Los Angeles is almost 2800.

Prefixes and Short Form Symbols

Before turning to "hands-on" metric work, let us look at some names that we have been using for metric sizes and also give their short form *symbols*. "Milli," "centi," and "kilo" are prefixes that we have used in connection with the word "meter," and these are prefixes that you will use again and again in the metric system because they are very handy for changing the size of a unit by a definite amount.

milli means "divided by a thousand"
centi means "divided by a hundred" and
kilo means "multiplied by a thousand"

Thus we began with the basic unit—the meter—and spoke of a "millimeter" to mean 1/1000 of a meter, "centimeter" to mean 1/100 of a meter, and "kilometer" to mean 1000 meters. And the same relations will hold true when we later speak of "milliliters" of volume or "kilograms" of weight. As to the short forms or symbols which are commonly used, we begin with the basic

---

[3] One notable exception is this: In architectural and engineering work, linear dimensions are always given in millimeters only. This is done so that nobody makes a mistake by putting a decimal point in the wrong place and creating a serious error. Thus, on the drawing of a steam locomotive the diameter of a pipe is shown as 45 mm, the diameter of a driving wheel 1400 mm, and the overall length of the locomotive 14,060 mm. However, it is easy to make a mental calculation to more familiar units and find that the wheel is 1.4 m in diameter and the locomotive very slightly over 14 m long.

ONE-KILOMETER DISTANCES
IN SELECTED MAJOR AMERICAN CITIES

Note: These distances were determined by scaling gasoline station highway maps and thus are not precise. However, they are sufficiently accurate to give a person a good mental image of a kilometer.

EAST

New York. From the west side of Central Park (8th Av.) look east across the park. Add one more block (to Madison Av.); the total distance is one kilometer.

Philadelphia. From the east side of City Hall Sq., look east along Market St. to Independence Mall; that distance is a bit more than one kilometer.

Washington. From the foot of the northwest steps of the Capitol Building, the straight line distance up Pennsylvania Av. to the National Archives Metro station is one kilometer.

Atlanta. The distance from the center of the Capitol Building facade south on Capitol Av. to an imaginary line from the northern point of Atlanta Stadium is one kilometer.

CENTRAL

Chicago. The distance from the foot of the north steps of the Field Museum straight north on Lake Shore Dr. to the center line (extended) of Buckingham Fountain is one kilometer.

St. Louis. In the park across Market St. from Union Station, stand opposite the west end of the station and look east to the Civil Courts Building. That distance is one kilometer.

New Orleans. From Canal St. (the beginning of the French Quarter) along Decatur St. to the far end of Jackson Square is one kilometer.

ONE-KILOMETER DISTANCES
IN SELECTED MAJOR AMERICAN CITIES
(continued)

WEST & SOUTHWEST

Denver. The distance from the east side of the State Capitol grounds (Grant St.) along Colfax (including the jog) to the west side of the U.S. Mint is one kilometer.

Dallas. From the Texas School Book Depository Building (also called the Dallas County Administration Building) along Elm St. to the Thanksgiving Tower is one kilometer.

Houston. The distance from the Sam Houston Coliseum (Bagby St. side) along Rusk St. to the far end of the Post Office is one kilometer.

PACIFIC

Seattle. The straight distance from the Space Needle east to the Times Building is one kilometer.

San Francisco. Golden Gate Park is somewhat less than one kilometer wide (north to south) for almost its entire length, but is one kilometer wide on the Stanyan Blvd. (or east) end.

Los Angeles. Exposition Park from Vermont Av. on the west to Figueroa St. on the east is a bit less than one kilometer long. Add one block (to Flower Dr.), and that makes one kilometer.

m which means "meter," followed by
mm which means "millimeter," then by
cm meaning "centimeter," and finally by
km to mean "kilometer."

Other symbols which we will encounter when we get to volume and weight will be formed in the same way. Notice that metric symbols are always written in small letters except when derived from a person's name such as W for Watt or V for Volt[a]. (Also there is one notable exception to be seen later at the top of page 28.) Symbols are not followed by a period as an *abbreviation* is, and are in the same form whether singular or plural, i.e., no "s" is used. Thus we write "1 m" or "15 m" (*not* "15 ms").

The next topic we will turn to is the "hands-on" use and learning to measure lengths in metric. By actually measuring things with metric tapes or rulers, you will come to understand how large (or small) things are in millimeters, centimeters, or meters. Kilometers, however, are more difficult to learn in the United States because long distances here are given on highway signs in miles only, and automobile speedometers/odometers (for any useful purpose) are calibrated only in miles. Using the picture of one kilometer in Chicago on the metric chart and using the list of One-Kilometer Distances in Selected Major American Cities will help, but the best way to learn kilometers is by repeatedly using kilometers.

If you go abroad, even to Canada or Mexico (but not to the British Isles), you are in a better position to learn kilometers. When you travel by highway, conscientiously watch the road signs and the speedometer/odometer. (Hopefully you are not in your American car calibrated in miles!) Notice repeatedly how many kilometers it is to the next town and to other towns; notice how many kilometers per hour the vehicle travels; and do not try to convert to miles. By being deliberately aware of these distances and speeds, you will find yourself becoming well adjusted to kilometers in about a week. In fact, the author knows this to be true because this is just how he learned kilometers while in Europe, and he did it in about a week.

Now to the "hands-on" use of most metric lengths. At this place in the course the instructor will divide the class into small groups for actual hands-on metric experience. This "lab session" will be followed by review problems to check on your learning and after that by a session on estimating metric lengths. Then comes a break prior to the section on Volume.

HANDS-ON EXERCISES

The fill-in-the-blank exercises below provide a useful introduction to comprehending metric measure. (All necessary equipment, namely, meter sticks, scales, liter pop bottles, and the like, have been furnished in advance by the instructor.) As you and your partner go through these exercises, be sure to pay attention to *just how long* so many meters or centimeters are; do not merely write down numbers without acquiring a concept of the size.

Bear in mind, you can read a book on how to play golf or how to play tennis, but you won't acquire any skill at playing golf or tennis unless you go out and actually play the game again and again. The same is true about learning metric measurement: you must simply use it to learn it. So, do not stop with these exercises. After you leave this class, *continue measuring sizes in metric as often as you can.* Not only will you learn metric, but also you will find that all your measuring will be much easier than it ever had been before.

(Sometimes it may be necessary to shorten this course. To do so, simply omit some of the exercises except for any of those marked with the symbol •.)

Dimensions Usually Given in Meters

Note: Record dimensions to two decimal places such as 7.85 m (which means 7 meters and 85 centimeters) or 0.93 m (i.e., no meters, just 93 centimeters); don't bother going the third decimal place (i.e., to millimeters).

1) The room in which I am is _____ m long, _____ m wide, and _____ m high.

• 2) A door in this room is _____ m high and _____ m wide.

• 3) The distance from the floor to the top of the door knob is _____ m.

4) A window in this room is _____ m high, _____ m wide, and is _____ m above the floor.

5) Here make a sketch, similar to the one shown, of a portion of the room where features such as doors and windows cluster together and write in the various dimensions.

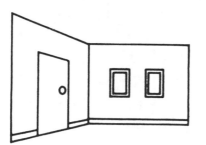

6) For further learning at home or in the workplace, make additional measurements such as those above including the size of an automobile, of the garage, etc.

If you have a camera with a focusing scale on its lens barrel, measure and note distances to various objects in meters, saying to yourself, "The distance to that tree is so many meters." This is a very convenient way to learn distances from about 0.5 m to 10 m (when using a 50 mm lens on a 35 mm camera).

An additional way to become familiar with some distances in meters is to keep in mind that:

The distance from the floor to the bottom of a man's belt buckle is about 1 m.

On a baseball diamond, the distance from home plate to second base is almost 40 m (actually, 38.8 m); so the distance from home plate to the pitcher's mound is about 20 m, and half that is 10 m.

Lengths and distances up to a kilometer are generally measured in meters.

Dimensions Usually Given in Centimeters
Note: For this exercise, record dimensions to the nearest centimeter, not to millimeters.

1) My height is _____ cm; my partner's height is _____ cm.

2) Ask your partner to make and record the following dimensions of your body:

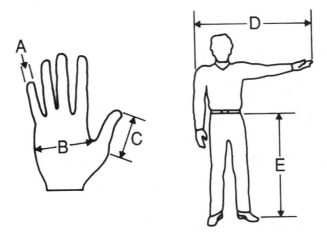

Notice that your body contains several built-in "measuring sticks" which are very close to the convenient sizes of 1 cm, 5 cm, 10 cm, and 1 m. For any approximate measurement, you always have your centimeter ruler handy!

Neck circumference (or shirt collar size) _____ cm.
Head circumference (or hat size) _____ cm.
Foot length (which, alas, is *not* shoe size) _____ cm.

3) Sketch furnishings in this room (chairs, desks, etc.) and write in the dimensions (such as floor to seat, height of a drawer, etc.) as measured.

4) In some other room (here or at home), make and record the dimensions of other centimeter-sized objects such as tables, lamps, TV sets, etc.

Item	Length	Width	Height
_____	_____ cm,	_____ cm,	_____ cm
_____	_____ cm,	_____ cm,	_____ cm
_____	_____ cm,	_____ cm,	_____ cm
_____	_____ cm,	_____ cm,	_____ cm

• 5) Books come in various sizes; some of them in this room are

Book A _____ cm high,   _____ cm wide,   _____ cm thick
Book B _____ cm high,   _____ cm wide,   _____ cm thick
Book C _____ cm high,   _____ cm wide,   _____ cm thick

Dimensions Usually Given in Millimeters

• 1) In the above books, measure the height of the letters.

Book A,   Cap ltrs _____ mm high,   Small ltrs _____ mm high
Book B,   Cap ltrs _____ mm high,   Small ltrs _____ mm high
Book C,   Cap ltrs _____ mm high,   Small ltrs _____ mm high

2) In a newspaper the capital letters are _____ mm high and the small letters are _____ mm high.

3) In a telephone directory the capital letters are _____ mm high and the small letters _____ mm high.

4) The dimensions of a wrist watch are (if rectangular) _____ mm long, _____ mm wide, _____ mm thick; the big hand is _____ mm long and the small hand _____ mm long.

If the watch is circular, its diameter is _____ mm, its thickness _____ mm, its big hand _____ mm long, and its small hand _____ mm long.

5) Because the contents of rooms (not to mention pockets or purses) are so different, it is impractical to try to name other specific objects which may be available. Here just choose, measure, and record whatever small objects may be at hand such as pencils, postage stamps, paper clips, door keys, and what-not.

Item   Dimensions

_____   _____

_____   _____

_____   _____

_____   _____

Distances Usually Given in Kilometers

Measuring distances in actual kilometers in this room, need we say, is scarcely feasible.   So, to try to get a concept of kilometers, apply the list of one-kilometer distances in various American cities which appears on pages 10-11. If you live in or are familiar with one or more of these cities your learning is rather easy; otherwise, just do the best you can.

Now, strange to say, acquiring a concept indoors of very long distances in kilometers is quite easy. Just perform the following exercise. However, because the exercise requires access to a globe of the earth and will take a bit of time, it could be inconvenient to perform in a class room environment. So perhaps it would be better to do this exercise elsewhere and on another occasion.

First take a long narrow strip of paper. The perforated strips from the sides of computer paper are excellent for this purpose. With Scotch tape, affix one end of the strip to the Equator on the globe and the other end to the North Pole. Mark these two reference points and remove the strip of paper. Recalling that a meter is defined as one ten-millionth the distance from the Equator to the North Pole, you realize that the distance from Pole to Equator is 10,000,000 m or 10,000 km (and, incidentally, that the circumference of the earth is 40,000 km). With the aid of a millimeter ruler, divide your strip into 10 equal parts; each of these represents 1000 km. Further divide the first part into four equal segments and you have 250 km subdivisions.

Using this paper scale, measure, record, and comprehend the straight line distances from

Chicago to
    New York _____ km
    Miami   _____ km
    Seattle   _____ km

New York to
    London _____ km
    Berlin _____ km
    Moscow_____ km

Seattle to
    Tokyo   _____ km
    Vladivostok _____ km
    Hong Kong _____ km

REVIEW PROBLEMS

Note: Twelve problems with multiple-choice answers follow. In each case at least one answer is correct and sometimes more than one. After working the problems and putting a circle around the letter of each answer which you consider correct, you can check your accuracy by using the answer sheet supplied at the end of this book. (Use a pencil for circling your answers instead of a pen so that you can correct any mistakes.) For your convenience in figuring, blank space for this purpose is provided on pp 22-25. The symbol • before a problem means that it must not be eliminated in making short versions of this course.

1) The basic unit in the metric system is the
    a) centimeter, b) millimeter, c) meter,   d) foot, e) kilometer

2) The length of a meter is
    a) the same as a Roman mile, b) 3 feet 8 1/4 inches, c) about the same as a city block, d) one ten-millionth the distance from the Equator to the North Pole, e) one degree of arc at the Equator

3) To visualize a meter quickly, all you need to do is imagine
    a) the height of a basketball player, b) the length of a fly swatter, c) the distance from your knee to the ground, d) the distance from the floor to a man's belt buckle, e) the distance from a man's shoulder to the end of his opposite outstretched arm

4) On a baseball diamond, the distance from home plate to the pitcher's mound is approximately how far?
    a) 10 cm, b) 1 m, c) 20 m, d) 45 m, e) 380 m

5) Visualize an ordinary 3-ring notebook. About how tall is it?
    a) 5 cm, b) 10 cm, c) 30 cm, d) 50 cm, e) 60 cm

6) Units smaller or larger than a meter are described by the prefix "milli," "centi," or "kilo" in front of the word "meter." Thus a "centimeter" is what part of a meter?

a) 1/1000, b) 1000 x, c) 1/100, d) 1/10, e) 10 x

7) The same question as 6), except that "a 'kilometer' is what part of a meter?"

a) 1/1000, b) 1000 x, c) 1/100, d) 1/10, e) 10 x

8) To measure the length (or height) of objects about the size of a human being or even considerably larger, we generally use meters. The length of very small objects is usually measured in

a) milligrams, b) kilometers, c) centimeters, d) millimeters, e) centiliters

9) In the metric system it is easy to go from one sized unit to another simply by moving the decimal point to the right or the left (unlike the English system in which moving a decimal point does not change feet to inches or miles). Thus, if a man is 1.66 meters tall, his height may also be expressed as

a) 166 mm, b) 166 km, c) 0.166 cm, d) 166 m, e) 166 cm

10) The best way to learn distances in kilometers is to

a) find the conversion factor and change to miles, b) use the scale on an American highway map, c) try to visualize multiples of 1000 meters, d) drive on a highway outside the United States and pay attention to road signs, e) learn approximate conversion factors

• 11) According to the survey which he got with the property, the length of a farmer's field is 4125 feet. The farmer wants to make a quick check of this figure and notices that his field lies beside a straight highway. So, to measure the length of the field, he simply drives his automobile from one end of the field to the other and observes on his odometer that the distance he drove was 0.8 miles. How well do the survey length and the measured length agree?

a) exactly, b) survey 93 feet longer, c) survey 76 feet longer, d) survey 6 feet shorter, e) survey 99 feet shorter

• 12) A French farmer in the same situation finds that the surveyed length of his field is 1817 meters. The odometer on his automobile gives a reading of 1.8 kilometers. How well do the survey length and the measured length agree?

a) exactly, b) survey 17 km longer, c) survey 17 m longer, d) survey 0.17 m shorter, e) survey 17 m shorter

ESTIMATIONS

Now that you have had some actual experience with metric in the Hands-on Exercises, you can apply and even improve this knowledge by estimating the sizes of various objects. While it is true that people usually cannot estimate sizes with a high degree of accuracy, it is also true that estimating is a helpful way to learn sizes. The reason is that estimating requires you first to visualize a standard size and then apply it to an object, thus embedding this size and its multiples in your mind. So in preparation, familiarize yourself with several standards such as a meter stick (a centimeter/millimeter ruler for shorter distances), a door practically 2 meters high, or an automobile 5 meters long and then mentally apply these standards when you do the following exercises.

As called for in the exercises below, estimate the various sizes, write them down, and then check your estimates by actual measurement. Finally, to determine just how well you conceptualize metric sizes, make the simple subtraction shown in the following example. As in golf, the lower your score (i.e., the smaller the number), the better your performance. (Why?)

Example: Mr. Jones and Mrs. Smith estimate and then measure the lengths of Items A, B, and C in meters. They write down the appropriate figures and then perform a simple calculation to see how accurate their estimates are, as follows:

Mr. Jones

Estimated	Measured	Difference	
Item A 3.75 m	3.0 m	Estimated	11.0 m
Item B 2.0 m	1.5 m	Measured	-9.0 m
Item C 5.25 m	4.5 m		+2.0 m
Total 11.0 m	9.0 m	(Conception of a meter is too short.)	

Mrs. Smith

Estimated	Measured	Difference	
Item A 2.5 m	3.0 m	Estimated	8.0 m
Item B 1.25 m	1.5 m	Measured	-9.0 m
Item C 4.25 m	4.5 m		-1.0 m
Total 8.0 m	9.0 m	(Conception of a meter is too long, but more accurate.)	

Now do the following exercises and, *most important,* make up and do many more such exercises as often as you can. Repeated estimation and use of metric will greatly improve your grasp of this measuring system.

Measurements Usually Made in Meters
Note: Try not to use the same room that was used in the previous Hands-on Exercises (p 12 ff).

		Estimated	Measured
1)	The length of this room is	_____ m,	_____ m
2)	The width of this room is	_____ m,	_____ m
3)	The height of this room is	_____ m,	_____ m
4)	The height of a door in this room is	_____ m,	_____ m
5)	The width of a door in this room is	_____ m,	_____ m
6)	The width of the hall outside this room is	_____ m,	_____ m
7)	The height of my partner is	_____ m,	_____ m

Choose two longer (probably outdoor) distances of about 10 to 50 m for 8) and 9).

8)	From _____ to _____	_____ m,	_____ m
9)	From _____ to _____	_____ m,	_____ m
	Total	_____ m,	_____ m

In the margin or on the following blank pages, determine how accurate your estimates are.

Measurements Usually Made in Centimeters

1)	The height of the chair seat from the floor	_____cm,	_____cm
2)	The width of the chair seat	_____cm,	_____cm
3)	The depth of the chair seat	_____cm,	_____cm
4)	The thickness of the chair seat	_____cm,	_____cm
5)	The width of this book is	_____cm,	_____cm
6)	The height of this book is	_____cm,	_____cm
7)	The height of my partner is (See 7 above. In borderline cases either m or cm may be used.)	_____cm,	_____ cm
	Total	_____cm,	_____cm

As above, check the accuracy of your estimates.

Measurements Usually Made in Millimeters

Note: Try not to use the same articles that you used in the Hands-on Exercises for Length.

1) The height of these CAPITAL LETTERS _____mm, _____mm

2) The height of these small letters _____mm, _____mm

3) The thickness of a pencil or pen _____mm, _____mm

4) The diameter of a coin (name it) _____mm, _____mm

5) The width of a wristwatch band _____mm, _____mm

6) The thickness of a door key _____mm, _____mm

7) The length of a door key _____mm, _____mm
   (As in the preceding 7, this too is a
   borderline case; either mm or cm
   are suitable.)   Total _____mm, _____mm

Again, check your accuracy.

Measurements Usually Made in Kilometers

As of the time of this writing, there is no convenient way to estimate and check kilometer distances in the United States. The best way to learn and to estimate distances and speeds in kilometers is to go abroad and drive an automobile (as described at the end of the section on Length, p 12).

But if you cannot go abroad, try this: Get the concept of a kilometer by using both the Chicago lakefront picture on the chart and the table of one-kilometer distances on pp 10-11 of this book. Then along some convenient straight street establish what seems to be one kilometer. If you are fortunate enough to have a long metric tape, you can check this estimate. Otherwise, you can use an approximate method based on the speed at which a person walks. An average adult walks at a speed of about 5 km/hr. So the distance that he/she walks in 12 minutes is close to one kilometer. Thus you can check your estimate just by walking and timing. Then make many other such estimates and check them also.

Use this blank page for recording information and for figuring.

Use this blank page for recording information and for figuring.

Use this blank page for recording information and for figuring.

Use this blank page for recording information and for figuring.

VOLUME

The objective of this section is to give students an understanding of metric units of volume and of their symbols (short forms).

"Volume" simply means "how much something holds." A box, for example, holds a certain amount, and its volume is determined by multiplying its length by its width by its height. Thus if a box is 40 centimeters long by 15 centimeters wide by 30 centimeters high, 40 x 15 x 30 gives 18,000 cubic centimeters as the volume. ("Cubic" refers to something that is the same length on each side—like an ice cube—and is used to measure volume even if, as in the present example, the box is not actually a cube.)

Cubic Centimeters, Milliliters, and Liters

Any volume can be measured either in cubic millimeters or in cubic centimeters. However, a cubic millimeter is so small that it is hardly ever used. A cubic centimeter is also small (just look at a metric ruler and imagine a box one centimeter on each side); nevertheless a cubic centimeter is often used and is very important in metric because it forms the basis of both the major unit of volume, the liter, and (as we will see in the next section) the basic unit of weight, the gram.

By definition, 1000 cubic centimeters make the major unit of volume, the liter. Most likely you already know how big a liter is because soda pop and many drinks today are sold in plastic one- or two-liter bottles. Hard liquor, too, is sold in metric sizes, with the one-liter bottle being quite common, as are the 750 mL and the 1.75 L bottles. Just look carefully at a one-liter bottle and you will know how big a liter is. Because the cubic centimeter is so small and the liter is of a convenient size, the liter is considered the standard unit of volume.

Now, with the liter as the standard unit of volume, let us divide it into 1000 equal parts. Recall from the section on Length that "milli" means "divided by a thousand," so these 1000 parts that we have of a liter are called "milliliters." In other words, *a milliliter and a cubic centimeter are the same thing.* It just depends on whether you are thinking first of a small volume and assembling 1000 of those to make a large volume, or whether you are first thinking of a large volume and then dividing it into small parts.

The liter along with the milliliter is the most commonly used unit of volume. In a physician's office people frequently have injections of, say, insulin in the range of 1 to 5 milliliters (or "cubic centimeters") and may have a blood sample drawn in the neighborhood of 10 to 20 milliliters. In the grocery store you often see fruit juices sold in 250 milliliter cartons or soda pop in 1- or 2-liter bottles. A coffee cup holds about 200 to 250 milliliters, and a normal-sized drinking glass around 250 to 300 milliliters. A small four cylinder automobile would have a gas tank holding about 40 liters, and a six cylinder car's tank would contain about 70 liters. Even very large volumes are expressed in liters. For example, in Germany houses are often heated by fuel oil, and a three story house would have in its basement an oil tank holding about 6 to 10,000 liters and sometimes more.

Cubic Meters

Very large volumes, however, are often expressed in cubic meters, that is, in units the size of a box that is one meter long on each side. Look at the picture on the metric chart of the girl visiting an aquarium to see how big a cubic meter is. Better, take a meter stick and form your own cubic meter.

Because a cubic meter consists of 1000 liters (go through the arithmetic yourself and you will see why), the 10,000 liter oil tank mentioned above could just as well be described as a 10 cubic meter tank.[4] For a much larger volume, consider one of those big things carried on a flat car in a freight train and called a "container." Its volume (as marked on the end of the container) is close to 68 cubic meters.

When we discussed length, we saw that some units are normally used when measuring small things and other units when measuring large things. Thus, millimeters and centimeters are used for short things, meters for longer things, and kilometers for long distances. The same is true in regard to volume; going from small to large we measure in cubic centimeters (or "milliliters"), then in liters, and finally in cubic meters.

$$\left.\begin{array}{l}1000 \text{ cubic centimeters} \\ \text{(or 1000 milliliters)}\end{array}\right\} = 1 \text{ liter}$$

$$1000 \text{ liters} \qquad\qquad = 1 \text{ cubic meter}$$

As to proper symbols for these units, they are just as easy as those for length, namely:

$$\left.\begin{array}{ll}\textbf{cm}^3 & \text{means cubic centimeter}^{5} \\ \textbf{mL} & \text{means milliliter}\end{array}\right\} \begin{array}{l}\text{Two names for} \\ \text{the same thing.}\end{array}$$

$$\begin{array}{ll}\textbf{L} & \text{means liter} \\ \textbf{m}^3 & \text{means cubic meter}\end{array}$$

---

[4] Granted, 10 m³ or 10,000 L may sound like too large a volume to be in a private home, but this statement is not incorrect. The author once lived in a three story house in Germany that had a 10,000 L tank and knew a family in another three story house with a 12,000 L tank.

[5] Among English-speaking people who regularly use metric (scientists and those in the medical field) "cc" (plural, "cc's") is by far the most commonly used short-cut term for "cubic centimeter." This form, although not approved by the General Conference on Weights and Measures, is so widely used that you must know it.

Remember we said in the section on Length that symbols for metric units are usually written in small letters with a few exceptions. One major exception is seen right here where capital "L" is preferably used for "liter" so that small "l" will not be confused with the number 1. Having said this, we must acknowledge that many people use "ml" instead of "mL." But common sense tells us that—because of confusion—this usage is to be avoided. Also we sometimes see a script "l," but that is certainly not as different from the number 1 as a capital "L" is. So don't use a script "l" either.

Now we reach the Hands-on Exercises for measuring volumes in metric, this to be followed by a set of Review Problems and then Estimations on Volume. After that comes a break before we go to our last topic, Weight.

HANDS-ON EXERCISES

The comments introducing the Hands-on Exercises for Length apply here also. Go back to page 12 and read them again. And as before, the symbol • designates exercises that must not be omitted in order to make short courses.

"Ordinary" Sized Volumes Are Stated in Liters

• 1) Look carefully at the 1 L plastic pop bottle at your seat. This will show you the size of a liter which is the fundamental metric unit of volume. Now examine some of the 2 L and 3 L bottles in the room; this will help you visualize larger volumes.

2) Look again at the 1 L bottle. You will notice that in addition to the liquid there is some air space. So which is a liter, the amount of liquid or the entire volume of the bottle? To find out, pour the liquid into a graduated 1 L or 500 mL kitchen measuring cup and record:
    The actual amount of beverage sold in a 1 L plastic pop bottle is _____ L.

    Now, using the same 1 L or 500 mL kitchen measuring cup, fill the bottle completely with water and record:
    The actual total volume of a 1 L plastic pop bottle is _____ L.

    So, when you buy 1 L of a beverage, do you get fair treatment or do you get cheated?

    Note: We can't put too much faith in these findings because a kitchen measuring cup cannot measure much closer than 25 or perhaps 50 mL. The basic idea, however, is reliable.

3) If there are enough 2 L and 3 L bottles in the room, repeat the exercise in number 2) with these larger bottles and record:

The actual amount of liquid sold in a 2 L bottle is _____L.
The actual volume of a 2 L bottle is _____ L.
The actual volume of liquid sold in a 3 L bottle is _____L.
The actual volume of a 3 L bottle is _____ L.

And again, note the warning following number 2).

4) An average adult man's total lung capacity is close to 5.8 L (a woman's is about 20% or 25% less). Combine the various bottles in such a way as to show these lung capacities.

• 5) An average adult person's body (whether male or female) contains close to 5 L of blood. As in the previous exercise, demonstrate about how much blood is in a man's or woman's body.

6) The total volume for air and fuel vapor in the cylinders of an automobile engine is called the "displacement" and is generally expressed in liters. Using the various containers on hand and filling them with the proper amounts of water, show the displacement of

     a 1.9 L engine
     a 2.2 L engine
     a 3.0 L engine
     a 3.5 L engine

Small Sized Volumes Are Stated in Milliliters (Cubic Centimeters)

1) The displacement of engines smaller than those used in automobiles, e.g., motorcycle or lawn mower engines, is given in cubic centimeters (cm^3, always called "cc's" in this connection). Remembering that a cm^3, a cc, and a mL are the same thing, show the displacement of

     a 900 cc (mL) motorcycle engine
     a 750 cc (mL) motorcycle engine
     a 170 cc (mL) lawn mower engine
     a 150 cc (mL) lawn mower engine

• 2) Drinking glasses come in various sizes. Those available in this room contain _____ mL, _____ mL, and _____ mL.

3) The fruit juice container at your seat is marked as containing 250 mL. According to the kitchen measuring cup, it contains _____ mL.

Again, remember that kitchen measuring cups cannot reliably measure volumes smaller than 25 or 50 mL. Scientific measuring equipment, of course, can measure much more accurately.

Large Volumes May Be Stated in Liters, But More Often in Cubic Meters

1) Using several meter sticks with your partner, show just how large a cubic meter is. (Remember that a cube is a box with each side the same length.) If you can demonstrate more than one cubic meter, so much the better.

2) Remember that the volume of a box is found by multiplying its length by its width by its height and that a meter is 100 cm long. Now calculate how many liters there are in a cubic meter, write the number here _____ L, and check with the answers at the bottom of this section.

3) The fuel oil tank for a three story house contains 10,000 L. This means it contains _____ m^3. (Check with the answer at the end of this set of exercises.) With your partner and with meter sticks, show how large this volume is.

    There are many different ways to attain this volume, such as having the tank
_____ m long, _____ m high, and _____ m wide, or
_____ m long, _____ m high, and _____ m wide, or
_____ m long, _____ m high, and _____ m wide, etc.

    (Because so many combinations are possible, only two combinations are given in the answers below.)

4) Take several cardboard boxes in this room and measure their dimensions.

Box 1 is _____ cm long, _____ cm wide, _____ cm high
Box 2 is _____ cm long, _____ cm wide, _____ cm high
Box 3 is _____ cm long, _____ cm wide, _____ cm high

    Now calculate their volumes and express them in cm^3, m^3, and L.

Box 1 contains _____ cm^3, or _____ m^3, or _____ L
Box 2 contains _____ cm^3, or _____ m^3, or _____ L
Box 3 contains _____ cm^3, or _____ m^3, or _____ L

    (In advance, the instructor has calculated the volumes of these boxes; he/she will give you the answers so that you can check yours.)

5) In the same way, calculate the volumes of the fish tanks or similar containers (if there are any) in this room and record their volumes in liters. Then, by using the measuring cup, fill the tanks with water (if any is conveniently available) one liter at a time. Record these data and see how well the calculated volumes agree with the measured volumes.

Tank 1: Volume calculated, _____ L, measured _____ L
Tank 2: Volume calculated, _____ L, measured _____ L
Tank 3: Volume calculated, _____ L, measured _____ L

Note: Very large volumes, such as those of railroad freight cars, diesel locomotive fuel tanks, city water tanks, etc., are usually given in cubic meters for "empty space" but in liters for liquids. Obviously, it is impossible to make such measurements here, but to give you an idea of sizes: An ordinary American box car contains close to the convenient number of 100 m³, a large American diesel freight locomotive carries 12,000 L of fuel oil, and a large tank for city water holds 2,000,000 L (often called 2000 m³).

Answers: No. 2) 1000 L; No. 3) 10 m³; then 5 x 2 x 1 m, or 2 x 3 x 1.6666 m, and so on.

REVIEW PROBLEMS

Circle the letter of the proper answer; it is possible for more than one answer to be correct. Again, • designates an item which must not be eliminated to make short versions of this course.

1) Although any volume could be expressed in cubic millimeters, these units are rarely used in everyday measurement because they
    a) are not a standard SI size, b) are not conveniently related to length, c) may be confused with ounces, d) are too small, e) cause confusion with decimal points

2) The cubic centimeter is an important unit of volume because it
    a) forms the basis of the major metric unit of volume (the liter) and the basic metric unit of weight (the gram), b) was computed directly from an astronomical unit, c) expands easily in a balloon, d) is of a convenient size for making ice cream, e) is much larger than a cubic foot

3) One liter is equal to
    a) 1000 mL, b) 100 cm³, c) 0.1 m³, d) 0.1 mL, e) 1000 cm³

4) A cubic centimeter and a milliliter are
    a) not much alike, b) large volumes, c) the same thing, d) not convenient for measuring gases, e) identical to a similar English unit

5) In American grocery stores today, a person often finds beverages sold in containers holding
    a) 1 L, b) 15 mL, c) 250 mL, d) 1000 mm³, e) 2 L

6) In an American liquor store today, the bottles most often found on the shelves contain
   a) 500 L, b) 1 L, c) 750 mL, d) 1.75 L, e) 3/4 gal

7) The typical capacity of the gas tank of a six cylinder automobile would be about
   a) 10 L, b) 20 L, c) 25 L, d) 70 L, e) 1200 L

8) Very large volumes are generally measured in
   a) square meters, b) cm^3 c) cubic meters, d) cubic millimeters, e) square kilometers

9) Because of the uniform naming pattern in the metric system, a cubic meter could also be described as a
   a) decimeter, b) deciliter, c) kiloliter, d) centimeter, e) centiliter

10) A drinking glass would most likely contain
   a) 250 mL, b) 0.3 L, c) 300 mL, d) 2.50 L, e) 8.50 mL

•11) The owner of a feed store is to receive a shipment of 10 bushels of bird seed which he will transfer into smaller boxes, each of which is 12" x 13" x 16 3/4". How many such boxes (to the nearest whole number of boxes) will he need to hold 10 bushels?
   a) 22, b) 27, c) 15, d) 9, e) 5.

• 12) A German feed store owner is expecting a shipment of 0.5 cubic meter of bird seed. He wants to do the same thing that the American store owner is doing and has boxes which measure 25 x 30 x 50 cm. To the nearest whole number, how many of these boxes will he need?
   a) 35, b) 141, c) 14, d) 45, e) 10.

ESTIMATIONS

The comments introducing the Estimations section for Length apply here also. Go back and read them again, but for standards of volume add a 500 or 1000 mL kitchen measuring cup, perhaps a 250 mL fruit juice container, plus a 1 L and a 2 L plastic bottle.

It is practically impossible to know just what containers of unknown volume would be available to various students, so this section cannot be specific. Instead, it must rely on students to furnish their own containers for estimating. Some such containers are wastebaskets, boxes, cups, drinking glasses, drawers, closets, and the like.

Although one can use cubic centimeters (or cubic meters) interchangeably with liters, "cubic" measurements are generally used for empty spaces and liters for liquids. This usage will be followed in these exercises on Volume. So now estimate volumes of your own choosing and, as in the Estimations of Length, check your accuracy. *Repeat such exercises* as often as possible to develop your concept of metric volumes.

To check your estimates of liquid volumes, pour water from the full container into a measuring cup and read the volume in cubic centimeters (the same as milliliters). To check empty space volumes, measure the container's length, width, and height in centimeters; multiply these figures together; the result is the volume in cubic centimeters.

Volumes Measured in Cubic Centimeters, Liters, or Cubic Meters
1) Try drinking glasses and cups of various sizes. For larger volumes try water buckets. Notice that the shape of a cup or glass (its sides may often be slanted or rounded) can fool you when making mental comparisons with a standard, straight-sided bottle or measuring cup.

2) Try whatever boxes, wastebaskets, or other containers that are available.

3) To estimate a large volume such as that of a closet or a room, you will use cubic meters. For a mental standard, just take a meter stick, go to a corner of this room, and lay out a cubic meter which you can then easily visualize.

Use the space here and on the following blank pages to do this work.

Use this blank page for recording information and for figuring.

Use this blank page for recording information and for figuring.

Use this blank page for recording information and for figuring.

MASS OR WEIGHT

The objective of this section is to give
students an understanding of metric units of
mass or weight and of their symbols (short forms).

Grams

What we usually call "weight" (when we actually should say "mass"[6])
is very simple to describe in metric because it is closely related to volume.
The weight of one cubic centimeter of water is simply taken as the basic unit
of weight and is called a gram. A gram is a very small amount and is almost
exactly the weight of an American dollar bill (or of a five, ten, twenty or
other bill for that matter). Often in restaurants you will find, on the table,
sugar-substitutes packaged in small envelopes of one gram each. Pick up one
of these or a dollar bill and see how light a gram is.

Kilograms

Because the gram is too light to be a convenient standard of weight, a
larger unit has been chosen. As you might expect, this larger unit is 1000
grams and, following the regular pattern of metric naming, is called a kilogram. Notice that 1000 grams of water occupy a volume of 1000 cubic centimeters or one liter. In the section on Volume, we said examine a plastic one
liter bottle of some beverage to see how big a liter is. Now we say, pick up
that bottle and you will know how heavy a kilogram is.[7] Simple and logical,
isn't it? Then go on with a full two liter plastic bottle and any combinations
of bottles to gain a concept of various metric weights.

---

6 To explain the difference between "mass" and "weight" would take too much space
and get us into complications. In everyday use we usually say "weight" when we
really mean "mass," and we all know what we are talking about; so there is no
need to do anything different here. In the fields of science, however, the distinction between weight and mass is important and is always made. Otherwise, only if
you are an astronaut experiencing "weightlessness" in outer space do you need to
be concerned about the difference between weight and mass. (In space you still
have your same mass but not your same weight!)

7 True, there are some errors in the above statements, but they are very slight. The
plastic one-liter bottle does weigh something—59 grams—but that is only a small
addition to 1000 grams. However, do not use a glass liter bottle; glass is much too
heavy. Also, there is a very slight difference in weight between equal volumes of
water, fruit juice, soda pop, beer, milk, and other drinks. For example, a liter of
apple juice weighs 1032 g, not 1000 g. But for everyday use in learning how heavy
a kilogram is, picking up a plastic one-liter bottle of some beverage is entirely
satisfactory.

Look carefully at the chart *All You Need To Know About The Metric System* and follow the development from length to volume to weight to impress this logical relationship on your mind. In so doing you will understand why we said early in this course that, in the metric system, everything including weight is based on that one fundamental unit of length, the meter. Now we will consider one more unit of weight closely identified with the meter, and that is the ton.

Tons

The previously discussed units of grams and kilograms are used to weigh everything but the heaviest objects. A nickel (that is, an American five cent coin) weighs 5 grams, and five nickels weigh 25 grams. European chocolate bars available in many American stores are commonly sold in 100 gram sizes, and the same is true of European soap available in select stores. A normal portion of meat at dinner is about 150 to 200 grams, and a grapefruit weighs around 500 grams. At birth a boy weighs about 3 kilograms[8], some 25 kilograms at 6 years of age, perhaps 60 kilograms at 20 years old, and 80 kilograms at 50 years. A football player weighs about 100 kilograms. Then a four cylinder American automobile weighs about 1000 kilograms and a six cylinder American car about 1500. And once we reach the "magic number" of 1000 kilograms, we have the largest unit of weight, the ton.

A metric ton is close to, but not the same as, the common (that is, the short) American ton. An American ton is equal to 907.2 kilograms compared to 1000 kilograms for the metric ton. Because of possible confusion between the two units, the metric ton is sometimes called by its French and German name the "tonne." But "tonne" looks strange to American eyes and pronunciation is uncertain, so we prefer to say simply "ton" or, if distinction is necessary, "metric ton," a term much more frequently used than "tonne."

Objects weighed in tons are just the ones you would expect. The automobiles which we cited as weighing 1000 or 1500 kilograms could just as well be said to weigh 1 or 1.5 tons. The railroad freight container referred to in the section on Volume (p 27) can hold 26.5 tons, and a very large diesel freight locomotive weighs 167 tons.

How is a ton closely related to a meter as mentioned above? You have probably guessed it already. One cubic meter (containing 1000 liters) of water weighs one ton, and this is clearly illustrated on the chart *All You Need To Know About The Metric System*. Examine this chart again to see how

---

[8] In countries where metric is the customary system of measurement, people use the short form "kilo" (in this case, pronounced "keelo") to mean "kilogram" just as we use the short form "photo" to mean "photograph." Thus in the above paragraph we would speak of a baby weighing "3 kilos" and an adult man weighing "80 kilos." The clipped form "kilo" refers only to weight and is never used to mean "kilometer," "kilowatt," or any other thousand-fold metric unit.

the various volumes and weights all fit together so nicely. Can you do the same thing with English units? Well, just try going from inches to cubic feet to gallons, pounds, and tons!

To set forth the relations among these units of weight and also to present their symbols we find that

The basic unit of mass or "weight" is 1 gram,
                    1000 grams = 1 kilogram,
            1000 kilograms = 1 ton.

                g means gram
                kg means kilogram
                t means ton

By finishing this section on Weight, we have covered all the material that we need to for a good understanding of metric for everyday use. Granted, there are other topics that we could consider such as land measure, pressure for automobile tires or for weather, food energy content, and whatnot. The list could go on, but we have done much work already and have certainly covered the most important aspects of metric for everyday use, so it is time to draw this instruction to a close.

And we near the end by turning to the last Hands-on Exercise, that of weighing objects in grams and kilograms. Naturally, we do not have any scales large enough to weigh in tons. After that the students will have the Review Problems on Weight and then the Estimations, to be followed by a short session on Temperature and a summary of the course.

HANDS-ON EXERCISES

Again, the introductory remarks to the Hands-on session for Length (p 12) apply here. Go back and re-read them before doing the following exercises. And as before, • identifies an item which must not be eliminated to make short versions of this course.

Light Articles Are Weighed in Grams
Note: In these exercises, weigh to the nearest gram or, if the scale is sufficiently sensitive, to the nearest 0.1 g.

1) An American dollar bill (or a five or a ten or any other) is said to weigh 1 g. Put such a bill on the scale and (if the scale is accurate enough) weigh the bill to the nearest 0.1 g. Write its weight here _____ g. Now put five bills on the scale and determine their average weight (i.e., the total weight divided by 5). It is _____ g. Let the dollar bill rest in the palm of your hand; notice how very light it is, that is, how light a gram is.

2) Put several paper clips on the scale and determine how many of them are necessary to make a gram: _____ paper clips.

     Now do the same by taking the average weight of a large number of paper clips, which gives _____ paper clips per gram.

3) Weigh a variety of nearby small objects (coins, keys, the 100 g chocolate bar in its wrapper, etc.); list their names and weights below:

Item	Weight	Item	Weight
_____	_____	_____	_____
_____	_____	_____	_____
_____	_____	_____	_____
_____	_____	_____	_____

4) An American nickel (i.e., a 5-cent coin) is said to weigh 5 g. Check this on a scale and write the measured weight here _____ g. Now take 5 nickels in the palm of your hand and be aware of the weight of 25 g.

     By combining other handy objects, form other rational-sized weights ("rational-sized" means quantities such as 50 g, 100 g, 200 g, 500 g, etc.) and hold each of them in your palm to get an idea of just how heavy these commonly used amounts are.

Most Articles Are Weighed in Kilograms

• *The following exercise is a basic and very important one.* It will show you the logical and convenient relationship between length, volume, and mass (or "weight") in the metric system. True, the units employed here are only those of volume and weight, but remember that volume is calculated from length (i.e., a cubic centimeter is the result of 1 cm x 1 cm x 1 cm).

•   1) Weigh an empty 1 L plastic bottle and record its weight here: _____ g.

     Now fill the bottle with a measured liter of water; weigh the filled bottle and record its weight here: _____ g.

     Subtract the weight of the empty bottle and see how close your result is to the expected 1000 g for a liter of water, then note that your result is _____ g heavy or light. (If your result is 1000 g, then it is 0.0 g heavy or light.)

     Again remember that a 1 L kitchen measuring cup is not a precision device, so you will probably not get the expected result, nor will you in the following step:

Divide the weight you obtained for a liter of water by 1000 to give you the weight of 1 mL (or 1 cm³) of water. Your result is _____ g, which is _____ g heavier or lighter than the expected 1 g.

A word of caution about interpreting the foregoing exercise: You must realize that the equality of one kilogram to one liter applies only to water. If you have a substance heavier than water—say iron, which is about 7.9 times heavier—then a liter of iron (actually, a 1000 cm³ piece of iron) would weigh 7.9 times as much or 7.9 kg. In the same way, a material lighter than water, such as wood, weighs less than the same volume of water (and that is why wood floats). A piece of elm weighs about 0.6 as much as water, so a 1000 cm³ block of elm weighs 0.6 kg.

Previously you held a dollar bill in your hand to see how light a gram was. Now take in your hand the plastic liter bottle full of water and (disregarding the slight weight of the bottle) realize how heavy a kilogram is. Hereafter you should have a good idea of the magnitude of the two fundamental metric weights, the gram and the kilogram. In conclusion, pick up full 1, 2, and 3 L bottles in various combinations so that you get a "feel" for 2, 3, 5, and other kilogram weights.

2) Weigh yourself while wearing normal indoor clothing. Your weight is _____ kg. Now remove your shoes and record your lighter weight as _____ kg.

From the exercise on Length (p 14), copy down your height here: _____ cm. Now you know your weight and height in metric.

3) Your lab partner, a man/woman who is _____ cm tall weighs _____ kg with shoes and _____ kg without shoes.

Give the foregoing data to the instructor who will record them and then tabulate the range of weights and heights of the people in this class. This will give you a good idea of what is meant by a small, a medium, or a large person.

4) Depending on what is available in the room, list some kilogram-sized objects and record their weights.

Item	Weight	Item	Weight
_____	_____	_____	_____
_____	_____	_____	_____
_____	_____	_____	_____
_____	_____	_____	_____

Very Heavy Articles Are Weighed in Tons

Adult men weigh around 75 to 80 kg. Very large men, such as football players, weigh about 100 kg and somewhat more. Heavier weights than these are beyond our realizing because we cannot lift them and thus "get a feel" for them. So when we get into tons—remember, a metric ton is 1000 kg, the weight of 1000 L of water or a cubic meter of water—we reach weights beyond our "feel." All we can say in this respect is that very heavy items such as automobiles, elephants, locomotives, shipments of coal, gravel, and the like are all weighed in tons. Naturally, we cannot weigh such things here.

REVIEW PROBLEMS

Circle the letter of the proper answers. As before, more than one answer to each question may be correct.

1) The basic unit of weight, the gram, is defined as the weight of
a) 1 L of air, b) 1 L of water, c) 1 cm^3 of air, d) 1000 cm^3 of air, e) 1 cm^3 of water

2) Approximately how heavy is a gram? The weight of
a) a flashlight battery, b) a dollar bill, c) a pencil, d) a small envelope of sugar substitute, e) ten paper clips

3) Because a gram is so very light, a larger unit has been chosen as the standard, namely, the kilogram. This unit weighs
a) 1000 cm, b) 1000 g, c) 1000 kg, d) 1000 t, e) 1000 lbs

4) One thousand grams of water occupy a volume of 1000 cm^3; thus we can also say that 1000 g of water occupy a volume of
a) 1 L, b) 1 m^3, c) 100 mm^3, d) 1000 mL, e) 1 cm^3

5) A good way to learn the weight of a kilogram is to pick up
a) an empty 1 L glass bottle, b) a 1 L glass bottle filled with water or some beverage, c) a usual bottle of beer, d) an empty 1 L plastic bottle, e) a 1 L plastic bottle filled with water or some beverage

6) Although we would not normally do so, we could say that a man weighs 0.083 tons. Because simply moving the decimal point to the right or the left in the metric system gives us the other units, we can say that—in usual terms—the man weighs
a) 83 g, b) 83,000,000 g, c) 8.3 kg, d) 830 kg, e) 83 kg.

7) The most common large unit of weight, the ton, is the weight of
    a) 100 m³ of water, b) 1000 cm³ of water, c) 1 m³ of water,
    d) 1000 L of water, e) 10 m³ of water

8) The most common large unit of weight, the ton, consists of
    a) 10 kg, b) 100 kg, c) 1000 g, d) 1000 kg, e) 1,000,000 g

9) A person wants to buy a water bed. But first he wants to know whether the bed will require too much water and thus weigh too much to be practical. The waterbed measures 4' x 7' x 9". How many gallons of water (to the nearest gallon) will it require to fill the bed completely?
    a) 157, b) 230, c) 56, d) 288, e) 62

10) How much will this quantity of water weigh?
    a) 362 lbs, b) 1435 lbs, c) 3/4 ton, d) 1311 lbs, e) 86 lbs

11) A man in Italy wants to do the same thing. The Italian waterbed measures 125 x 200 x 25 cm. How many liters of water (to the nearest liter) will it contain?
    a) 256, b) 600, c) 315, d) 625, e) 1500

12) How much will this quantity of water weigh?
    a) 256 kg, b) 6.00 kg, c) 3150 kg, d) 62.5 kg, e) 625 kg

ESTIMATIONS

The comments introducing the Estimations section for Length apply here also. Go back and read them again. For standards of weight use a dollar bill for a gram and a nickel for 5 g (also try to make 25 or even 50 g standards). Hundred gram soap bars and 100 g European chocolate bars are available in some stores and make convenient standards; try to get some of them for weights of several hundred grams. Then, for a kilogram (1000 g) use a 1 L plastic bottle filled with some beverage; several of these will give you even more kilograms. (Note: The weight of the paper wrapper on the soap or the chocolate as well as the weight of the plastic bottle is quite small and can be ignored for the present use.)

Do the following exercises and check your accuracy by using both a gram scale and a kilogram scale furnished by the instructor. And after you have finished these exercises, be sure to do others of the same kind later on.

Light Weights Are Usually in Grams

	Estimated	Measured

1) The weight of a paper clip _____ g, _____ g

2) The weight of a pencil (or pen) _____ g, _____ g

3) The weight of a shoe _____ g, _____ g

4) The weight of an apple _____ g, _____ g

5) The weight of a can of food _____ g, _____ g

6) The weight of a medium sized book _____ g, _____ g

7) The weight of a large book _____ g, _____ g

Total _____ g, _____ g

As in the section on estimating Lengths, determine how accurate your estimates are.

Most Weights Are Usually in Kilograms

Note: Because there is no way to know what items you have conveniently at hand, you are to supply your own items in the blank spaces.

1) The weight of several apples _____ kg, _____ kg

2) The weight of .... _____ kg, _____ kg

3) The weight of .... _____ kg, _____ kg

4) The weight of .... _____ kg, _____ kg

5) The weight of .... _____ kg, _____ kg

6) The weight of .... _____ kg, _____ kg

7) The weight of a large book
(See 7 above. In borderline cases
either g or kg may be used.) _____ kg, _____ kg

Total _____ kg, _____ kg

As above, check the accuracy of your estimates.

Large Weights Are in Kilograms, and Very Large Weights in Tons

An average adult man weighs about 75 or 80 kg and a small 4 cylinder automobile about a ton. It is very difficult or perhaps impossible for most people to lift a man, and nobody can lift a ton, so we have no way to "get a feel" for the weight of such objects, let alone of heavier ones.

The following blank space is for your own figuring.

Use this blank page for recording information and for figuring.

Use this blank page for recording information and for figuring.

Use this blank page for recording information and for figuring.

TEMPERATURE

The objective of this section is
to learn temperatures in Celsius.

A S A POSTSCRIPT we will consider temperature. The temperature scale used with metric measurements, it is interesting to note, had already been in existence for about 50 years before French scientists of the 1790s created metric. What they did in regard to temperature was simply to adopt this pre-existing centigrade (now called "Celsius") scale.

Like metric, Celsius is sensibly based on 100. Water freezes at 0 degrees and boils at 100 degrees (written 100 °C with a space between the figure and the symbol °C)[9], and on this scale normal body temperature turns out to be an even 37 °C. The way to learn Celsius temperature is simply to use it, and fortunately this can be quite easily done in the United States. In our cities, most large outdoor thermometers give temperatures alternately in degrees Fahrenheit and in degrees Celsius. So the thing to do is to pay attention to the Celsius figure, ignore the Fahrenheit (which can lure you into the trap of conversion!), and simply be aware of what the surrounding air feels like. If the thermometer reads 20 °C, you will find that the environment is very comfortable. Then go and find another thermometer which is in the shade or in the sun. There you may find the temperature to be 16 °C or 24 °C; in both cases pay attention to how chilly or warm you feel and, by repeatedly doing such an exercise, you will find that in about a week you are well adapted to Celsius temperature. (As in the previously-cited case of learning kilometers, this was the author's personal experience in Europe. It can be yours, too, and right here in the United States.)

As to the range of Celsius temperature which you will encounter, the coldest it ever gets in Chicago is about –30 °C, which is very cold, and the hottest is around +40 °C, which is very hot. Thirty degrees is hot, 25 °C is warm, 20 °C is most comfortable, 10 °C is chilly, 0 °C is the freezing point of water, –10 °C is cold, –20 °C is very cold, and –30 °C is extremely cold.

It's easy to become accustomed to Celsius temperature, and it fits very well with metric measurement.

---

[9] However, if it is understood that the temperature scale is Celsius, then you simply write 100° with no space and no C. "Celsius" refers to the Swedish scientist Anders Celsius (1701-1744) who is credited with devising this temperature scale.

CONVERTING FROM ONE SYSTEM TO ANOTHER

The objective of this section is to learn
how to convert from English to metric
measurement and vice versa when necessary.

VERY EARLY IN this course we said *Don't convert!* And that sound advice remains unchanged while you are learning metric. Converting will simply cause all sorts of nuisance and, quite possibly, prevent you from learning metric because you will always be going back to English units anyway. So, *Don't convert.*

Having said that—nay, having insisted on that—we get to the place where we tell you how to convert from one system to the other! This apparent contradiction is not as serious as it looks. After a person has learned metric and feels at home in it (you are probably not there yet, so please practice more and get there), then he/she can convert from one system to the other if necessary. And with the United States having used English units for so long a time and with metric now coming into use, you know that there will be many occasions when conversion is necessary.

At your place of work, for example, you may have shelves that are 10 inches apart and the new products which you have ordered are 30 centimeters high. Will they fit? Your doors are 30 inches wide and you are expecting the delivery of some boxes which are 1.5 meters wide. Will they go through the doors? A machinist receives a drawing dimensioned in millimeters, and his tools read only in inches. Now what?

These and other such problems will frequently be encountered as the United States changes to metric. However, with the use of a hand calculator or a proper computer software program (such as "S.I. Plus" from Geocomp Corporation or "Metric-X" from Orion Development Corporation) it is not difficult to change from one system to the other. Some hand calculators (especially those available from Blackhawk Metric Supply) are made specifically for metric conversion and give you the right answer directly. Moreover, you can take them with you anywhere. With a computer conversion program, of course, you are limited to the location of the computer, but often that is no handicap. If, however, you use an ordinary hand calculator, then use the conversion factors on the next page which will cover your most common needs. For conversions not shown there, it will be necessary to get a more complete conversion table. A big dictionary is a likely source of such a table, and an especially complete table is to be found in the most recent edition of *International System of Units (SI): The Modernized Metric System*. This publication is available from the American Society for Testing and Materials, 100 Barr Harbor Dr., West Conshohocken, PA 19428; telephone 610-832-9500; FAX 610-832-9555; e-mail service@astm.org

If you know	And want to find	Multiply by
inches	centimeters	2.540
feet	meters	0.305
cubic inches	cubic centimeters	16.387
cubic feet	cubic meters	0.028
dry quarts	liters	1.101
liquid ounces	milliliters	29.574
liquid quarts	liters	0.946
ounces (weight)	grams	28.350
pounds	kilograms	0.454
tons (2000 lb)	tons (metric)	0.907

If you know	And want to find	Multiply by
centimeters	inches	0.394
meters	feet	3.281
cubic centimeters	cubic inches	0.061
cubic meters	cubic feet	35.314
liters	dry quarts	0.908
milliliters	liquid ounces	0.034
liters	liquid quarts	1.057
grams	ounces (weight)	0.035
kilograms	pounds	2.205
tons (metric)	tons (2000 lb)	1.102

SUMMARY OF METRIC

THE METRIC SYSTEM of measurement is easy to understand and simple to use. Its main unit is the *meter*, originally defined as one ten-millionth the distance on the earth's surface between the Equator and the North Pole. Every other common unit, whether for length, volume (capacity), or mass (weight), is based on the meter. These other units, if small, are related to a meter as 1/10th, 1/100th, or 1/000th; or, if large, are 10, 100, or 1000 times bigger.

In the metric system, you will often see three handy words which tell how big a new unit is when compared to a base unit such as "meter," "liter," or "gram." These words (or "prefixes") are:

milli which means "divided by 1000" as in "millimeter"
centi which means "divided by 100" as in "centimeter"
kilo which means "multiplied by 1000" as in "kilometer"

There are many other prefixes which take you by steps of 10 to both extremely large and extremely small sizes. These are used in scientific work but are not often encountered in daily life, so that is why they are not listed here. However, if you do come across an unfamiliar prefix, you will probably find someone who can explain its meaning and use to you. Moreover, you can find a complete list of metric prefixes in many recently-published large dictionaries.

The symbols and the names of the most common metric units follow. Notice that, in writing, no "s" is added to the symbol even for plural use, thus: 1 mL, 75 mL (*not* 75 mLs).

Units of Length

mm = millimeter
cm = centimeter
m = meter
km = kilometer

Units of Volume (Capacity)

cm³ = cubic centimeter ⎫ These are two names
mL = milliliter ⎬ for the same thing.
L = liter
m³ = cubic meter

Units of Mass (Weight)

g = gram
kg = kilogram
t = ton

TECHNOLOGICAL SUPPLEMENT

TECHNOLOGICAL SUPPLEMENT

INTRODUCTION

*M*ETRIC FOR ME! was written for laypeople. However, during the short time that it has been on the market, many copies have been bought by people in various fields of technology. So, to serve future technical readers better, the author has been pleased to write this brief TECHNOLOGICAL SUPPLEMENT. The SUPPLEMENT presents and briefly explains some basic technological units pertaining to **Mass, Weight, Force, Pressure, Heat, Energy, Work, Power,** and **Torque**. The SUPPLEMENT does not try to teach engineering. On the contrary, any engineer or other technical person who uses this book and SUPPLEMENT presumably already knows his/her own field and how to go about solving problems in it. What the SUPPLEMENT does is introduce the person to those basic metric units to be used in his work instead of English units. Once the engineer understands key elements of technological metric units, he/she can then carry on alone.

Two North Central College colleagues of the author's are to be thanked for their critical readings of the following material and for their helpful comments thereon. These two are Dr. David A. Horner, Professor of Chemistry and Physics, and Dr. Paul W. Sutton, Harold and Eva White Distinguished Professor in the Liberal Arts and Professor of Chemistry and Computer Science.

SOME EVERYDAY UNITS

Area

The area of rooms, buildings, city lots, and the like, is measured in square meters. The unit of measure for large parcels of urban land and for farm land is the *hectare*, a rectangular area comprising 10,000 m^2 (for example, 100 x 100 m). A hectare (also spelled he*k*tare) is easy to visualize. Imagine an American football field with only one end zone; the overall length is close to 100 m. Now make a square out of this. The resulting square is one hectare. Larger areas such as those of state parks are expressed in square kilometers.

Automobile Fuel Efficiency

In the metric system, automobile fuel efficiency is expressed in liters per hundred kilometers (L/100 km), a very convenient measure. For example, if your VW Golf has a 55 L gas tank and uses 7.5 L/100km, dividing 55 by 7.5 shows that you can go 733 km on a tank of gas. Thus, if you're driving from Chicago to Cleveland, ca 500 km, or even an additional 100 km to the Pennsylvania state line, a full tank of gas is more than adequate. But if you're thinking of driving to Altoona, Pennsylvania, an 800 km trip—look out! Or if you're using your big Mercedes 500 SL with an 80 L tank and a fuel consumption of 13 L/100 km, you can drive only 615 km and will be lucky to reach Pennsylvania.

SOME BASIC UNITS IN PHYSICS

Mass, Weight, and Force

Now we must distinguish between "mass" and "weight." *Mass* is an innate physical property of any substance, the property that gives a substance its inertia or its resistance to acceleration. *Weight* is a *force*, and force is defined as mass times acceleration, or $F = m \cdot a$. Weight is defined as the special case of force when a = the acceleration due to the attraction of gravity at the surface of the earth, averaging 9.81 m/s^2. Thus, if an object has a mass of 1 kg, substitution in the formula shows the weight to be 9.81 kg·m/s^2. But since, by definition, a force which imparts to a mass of 1 kg an acceleration of 1 m/s^2 is called a *newton* (not capitalized; but symbol is capital N), we see that something with a *mass of 1 kg has a weight of 9.81 N.*

In everyday life we do not distinguish between weight and mass. We never say that a man has a mass of (or that "he masses"!) 85 kg and weighs 833.85 N; we incorrectly say he weighs 85 kg, and everyone knows what we mean. In science and engineering, however, the situation is different. If, for example, in a lunar exploration project the difference between mass and weight were ignored, the calculations involving weight would have about a six-fold error because lunar gravity is approximately one-sixth of that on earth. The above person's mass would still be 85 kg, but his weight would be about one-sixth his weight on earth or approximately 140 N.

To get "a feel" for magnitudes in newtons: Hold a small apple, its mass about 100 g, in your hand. The downward force that it exerts on your hand (its weight) is about 1 N, while a football player with a mass of 102 kg exerts a downward force of (or weighs) 1000.62 N—very close to 1 kN (1 kilonewton or 1000 newtons). Then the thrust of a jumbo jet airplane engine is about 1 MN or 1 meganewton, i.e., 1,000,000 newtons.

Pressure

From force we go directly to *pressure* which is simply force per unit area and is expressed in *pascals* (Pa), a pascal being one newton per square meter or, symbolically, 1 Pa = 1 N/m^2. Because this is a small unit amounting to the downward force of an apple on a square meter, we generally employ a unit one thousand times larger, the kilopascal (kPa) or 1000 N/m^2. By way of illustration, the tire pressure for a small automobile (such as a VW) is about 200 kPa. But although pascals were designated as the official SI metric units of pressure by the international General Conference on Weights and Measures in 1960, older pressure units continue to be the most commonly used ones in many parts of the world. Therefore, they must be learned and understood.

These older units are the *atmosphere*, the *bar*, the *millibar*, and (for vacuum) the *torr*. One atmosphere is defined as "normal" air pressure at sea level, or the pressure which will support a column of mercury 760 mm high[10]. The tire pressure in the above-mentioned small automobile is close to 2 atm (virtually 200 kPa) while the pressure in the tire of a large, six-wheeled bus is about 8 atm (or 800 kPa). Then the pressure in the boiler of a steam locomotive used to haul passenger trains is around 18 atm. One distinct merit of the atmosphere as a unit of pressure is that it is intuitively understood by almost everyone, thus its continued use.

Bars and millibars are the units employed by professional meteorologists the world around (including the USA) to measure air pressure. Likewise in their weather reports to the general public, the world news media (except the American and a few others) give air pressure in bars and millibars (though sometimes in millimeters of mercury). Normal pressure at sea level is close to one bar or 1000 millibars, practically the same as one atmosphere. In Chicago, over the course of a year, the air pressure hovers around 1000 mbar and ranges from about 990 to 1045 mbar. The torr, used for measuring vacuums, is defined as one millimeter of mercury.

The *bar almost equals one atmosphere*—but not quite. And here we encounter the only conspicuous "glitch" in the metric system known to this author. A bar is defined as equaling 100 kPa. But an atmosphere, instead of also being a nice 100 kPa, turns out to be 101.325 kPa or 1.01325 bars at the "normal" air pressure of 760 mm of mercury. However, at the more rational value of 750 mm of mercury[11], a bar is only 0.019,345 kPa different from (less than) an atmosphere, and many barometer scales have a distinct marking at the coinciding 750 mm/1 bar position. This author has never discovered the explanation for these annoying inconsistencies, but he knows that going any further into them here would only lead into non-rewarding complexity.[12] Fortunately, though, "these differences in values," according to Oxford University Professor P.W. Atkins, "are usually negligible except in work of the highest precision."[13] Thus for everyday use we can say 1 atm = 1 bar = 100 kPa at 750 mm and can employ these units interchangeably. But for unquestionable accuracy, scientists, engineers, and people in related fields must do their work in kilopascals and use 101.325 kPa (760mm) as normal air pressure.

----

[10] Notice the odd practice of measuring pressure—a force—in units of length. But this has been done ever since the 1640s when Evangelista Torricelli (after whom the torr is named) did his pioneering work in barometry. The modern units of pascal and kilopascal, however, are true units of force.

[11] Actually, the value is 750.062!

[12] The situation is no better in the English system where normal air pressure, instead of being an even 30 inches of mercury, is 29.92.

[13] P[eter] W[illiam] Atkins, *Physical Chemistry*, 5th ed (New York: W.H. Freeman and Co., 1994), p 77, n 2.

Heat, Energy, and Work

Now by way of another uncertainty, albeit a minor one, we return to the realm of certainty in measurement. We go from heat as measured by the *calorie*—familiar to every American at least by name—to the realm of *energy* and *work*, the *joule* and the *watt*.

The *calorie* (cal) is simply defined as the amount of heat necessary to raise the temperature of one gram of water one degree Celsius. However, this concept was formulated before the relationship between heat and work was clearly understood and since then has been replaced in the SI version of metric by the *joule* (J) which is defined as one newton meter, or 1 J = 1 N·m. Moreover, the joule is also the fundamental unit of *both energy and work*.

To look a bit more closely: The joule is a measure of energy, including the molecular kinetic energy in a substance, and kinetic energy is defined as one half the mass times the square of the velocity (here, of particles on the molecular level) or

$$KE = 1/2 \ m \cdot v^2.$$

Now, work is defined as force times distance, or

$$work = F \cdot d.$$

But since it is also true that $F = m \cdot a$, we can substitute and get

$$work = (m \cdot a)d.$$

If mass is expressed in kg, acceleration in m/s^2, and distance in m, then

$$work = kg \cdot m^2/s^2.$$

Remembering that a newton equals a kilogram meter per second squared and that a joule equals a newton meter, then

$$1 \ J = (1 \ kg \cdot m/s^2)m \ or$$

$$kg \cdot m^2/s^2, \text{ the same as work.}$$

Thus we see why the joule is the fundamental unit of both energy and work.

In terms of the familiar calorie, 1 cal = 4.184 J, meaning that 4.184 J are necessary to raise the temperature of a gram of water by 1 °C or that 4,184 J are necessary to raise the temperature of a kilogram (liter) of water 1 °C.

The joule is very useful because it is a unit of *any* kind of energy, be it work, heat, kinetic, potential, or what-have-you. As such, the joule replaces all other energy units whether calories, foot pounds, ergs, Btu's, and so on, over fifty in all! The simplification of technical work which results from one unit taking the place of some fifty previous units is quite apparent.

Power and Torque

Our next topic, *power*, is defined as the time rate of energy consumption or generation. If one joule of energy is consumed or generated in a second, the result is called one *watt* (W). That is, a joule equals a watt second, or

$$1 \text{ J} = 1 \text{ W·s} \text{ or (as seen earlier) } 1 \text{ J} = 1 \text{ N·m}.$$

The N·m is also the unit for *torque*. But be careful. Torque and energy are not the same, and torque is not measured in joules but only in newton meters. Torque and energy, although dimensionally the same, represent different concepts. Torque is a vector quantity which has both magnitude and direction; and energy is a scalar quantity which has magnitude only.

As the joule is replacing other units of energy, so the watt is replacing other units of power such as the horsepower, the cal/sec, Btu/sec, etc. For example, a German automobile publication designates a Chrysler LeBaron with a 2.2 L engine as a 94 hp or 69 kW car, a Ford Taurus 3.0 L as a 142 hp or 104 kW car, and a Buick 3.8 L Electra as 167 hp or 123 kW auto. (Alas, there are six or more different kinds of horsepower in use. In this case, the magazine was using 1 metric hp which is equal to 0.7355 kW.)

Now that we are dealing with watts, we are in the realm of customary *electrical* units, and for Americans it is nice to know that, when working with electricity, they have always been using metric units and need to learn nothing new! Watts, kilowatts, amperes, ohms, and the like are the same ones that Americans have been using all their lives.

As to the magnitude of various power units, we all have an understanding of a 60 or a 100 W light bulb. A flashlight bulb is 1 W, a two-horse team has approximately 1.5 kW of power and, as just seen, medium-sized automobiles have a little over 100 kW of power.

There are still more units that could be considered, e.g., those used in flow, in viscosity, or in luminous intensity. But to treat all or even many of them would extend this SUPPLEMENT into a long coverage, which is not its purpose. Rather, this SUPPLEMENT introduces the reader to some of the most important units for technical purposes. By following the pattern of how he/she learned these units and how to apply them, the engineer or technician can hereafter learn other units particular to his/her field as needed.

ANSWERS TO REVIEW PROBLEMS

LENGTH (p 17)	VOLUME (p 31)	WEIGHT (p 42)
1) c	d	e
2) d	a	b,d
3) d,e	a,e	b
4) c	c	a, d
5) c	a,c,e	e
6) c	b,c,d	e
7) b	d	c,d
8) d	c	d,e
9) e	c	a
10) d	a,b,c	d
11) e	d	d
12) c	c	e

DAMAGE
NOTED

FREEPORT MEMORIAL LIBRARY

3 1489 00431 3571

530.812   Shoemaker, Robert W.
S
         Metric for me!

DATE			

17.00   6-28-00

FREEPORT MEMORIAL LIBRARY
Merrick Road & Ocean Avenue
Freeport, N. Y. 11520

BAKER & TAYLOR

FREEPORT MEMORIAL LIBRARY